HOW TO START GOAT FARMING BUSINESS

A complete guide on raising goats

for beginners

Steve K. Bryant

COPYRIGHT

CONTENTS

INTRODUCTION

Describe and Extend Goat Farming

Breeding, raising, and managing domestic goats (Capra aegagrushircus) is referred to as goat farming, or caprine farming. It is a centuries-old discipline of animal husbandry that makes a substantial global contribution to the agricultural industry. The production of meat, milk, fiber (such as cashmere and mohair), and even animals for pets or enjoyment are all included in the diverse components of goat farming.

Goat farming has a broad and flexible application, with several breeds suitable for particular uses. Certain varieties are highly valued for their luxuriant fiber or high milk yield, while others specialize in producing meat, making them a great source of protein. Goats are also renowned for their resilience and capacity to flourish in a variety of settings, including mountainous and desert locations.

Goat farming's significance for animal husbandry

In animal husbandry, goat farming is very important for a number of reasons:

Sustainable Livelihoods: Millions of people worldwide, especially in rural and marginalized communities, rely on goat farming as a source of income and a means of subsistence. It offers small-scale farmers the chance to make money by selling goat products including meat, milk, and fiber.

Nutritional Security: The meat and milk produced by goats are excellent sources of premium protein. Their capacity to transform subpar fodder into wholesome sustenance renders them especially valuable in areas with restricted availability of other protein sources.

Environmental Benefits: Goats help with sustainability and land management. Their grazing habits aid in regulating the growth of the flora, lowering the chance of wildfires and enhancing biodiversity. Goat dung also acts as an organic fertilizer, increasing agricultural yield and improving soil fertility.

Adaptability: Goats are hardy creatures who can survive in unforgiving conditions with little food. Their resilience to harsh weather conditions, such as heat waves, droughts, and subpar fodder, makes them invaluable tools for farmers.

Cultural and Social value: Goats are frequently used in customary rites, festivals, and ceremonies, giving them cultural and social value in many civilizations. Goat farming also promotes empowerment and communal cohesion, especially among women and underrepresented groups.

In conclusion, goat farming is essential for advancing sustainable development, food security, and the reduction of poverty. Its diverse benefits to livelihoods, nutrition, and agriculture highlight its importance in the field of animal husbandry. Upon delving more into the complexities of productive goat husbandry, it is apparent that raising these adaptable animals has enormous benefits for people, societies, and businesses.

CHAPTER ONE
How to Begin Farming Goats

Planning and thoughtful thought are essential when beginning a profitable goat farming operation. The necessary measures to start your path into goat farming are outlined in this chapter.

Choosing Goats with Good Breeds

The success of your goat farm depends on your choice of breed. Take into account elements like:

Determine: Decide if your goal is to raise goats for pets, meat, milk, or fiber.

Breed Characteristics: Every breed has unique qualities that make it suitable for different environments and uses.

Genetics and Health: Select goats from reliable breeders that have a track record of producing healthy, superior animals.

Local Adaptability: Choose breeds that are suited to your particular climate and geographic area.

Giving Adequate Nutrition and Care

Your goats' productivity and well-being depend on providing them with the right nourishment and care. These are important points to remember:

Fresh food and clean water

Water: Always have access to fresh, clean water. Goats need to drink a lot of water for good digestion and general wellness. **Food:** Provide a well-balanced diet of premium forage, like grass and hay, with grains, minerals, and vitamins added as needed. Make sure there are no pollutants or mold in the feed.

Observance of Expectant Mothers and Children

Prenatal Care: During the gestation period, offer sufficient nutrition and veterinary care while keeping a tight eye on expectant mothers.

Joking: Set up a tidy and secure space for jokes. If needed, offer assistance throughout childbirth and postpartum care for the mother and her children.

Child Care: To boost immunity, make sure newborns are given colostrum within the first few hours of life. To aid in their growth and development, give them the proper food, warmth, and shelter.

The Necessary Goat Count for Novices Goats are gregarious creatures that do well in groups. To avoid loneliness and encourage social contact, it's best to begin with a minimum number of goats if you're just starting off. Take into account these recommendations:

Minimum Herd Size: To avoid boredom and loneliness, try to start with at least two goats. Appropriate choices are a buck and a doe, two does, or a doe and a wether.

Dynamics of Herds: Bigger herds need more room and resources, but they also present more breeding opportunities and diversity. Considering your experience level and available resources, start with a reasonable amount. Establishing a suitable herd size at the outset, giving adequate care and nourishment, and choosing well-bred goats are all necessary for a prosperous goat farming endeavor. The long-term viability and sustainability of your farm will depend on your ongoing education and adaption to your goats' demands.

CHAPTER TWO
What Makes Goat Growth Successful

Several aspects that affect the animals' growth and development are critical to the success of goat farming. Goat health and productivity are contingent upon the comprehension and optimization of these variables. We examine three major factors that greatly influence goat growth in this chapter: preserving the quantity and quality of feed, establishing the best possible growth and development environment, and appreciating the value of social contact and herd size.

Preserving the Quantity and Quality of Forage

As inherently foraging animals, goats consume a wide variety of foods, such as grasses, bushes, and other plants. Enough availability to high-quality fodder is necessary for their general health and growth. Every day, goats usually eat

1.5% to 2% of their body weight in pasture. To maintain ideal growth rates, it is crucial to guarantee a steady supply of nutrient-rich fodder.

Effective grazing management techniques should be used by goat farmers in order to maintain the quantity and quality of feed. Goats that are rotated to new pasture areas on a regular basis are able to avoid overgrazing and guarantee enough regrowth of the fodder. Hay or other supplemental feeds can also be added to grazing to help meet nutritional needs, particularly in the winter months when forage supply is restricted.

It is crucial to routinely check the quality of the feed and the conditions of the pasture. It is imperative for farmers to consider variables including the mix of plant species, maturation stage, and nutrient content in order to guarantee that goats receive optimal nutrition. Farmers may encourage the best possible growth and development in their goat herds by providing them with high-quality fodder.

Optimal Conditions for Development and Growth

In goat farming, it's crucial to establish the right environment for growth and development. Environments that provide sufficient shelter, uncontaminated water, and hygienic conditions are ideal for goats. Ensuring a cozy and stress-free living space is essential to optimizing growth rates and reducing health problems.

Goats should be housed in facilities that are suitable for preventing disease spread, extreme weather, and predators. To support health and well-being, adequate ventilation, bedding, and space allowance are crucial factors.

Herd health management strategies should be given priority by goat farmers in addition to physical infrastructure. This covers routine immunizations, deworming, and health-promotion activities. Goats in good health have higher development potential and are more resistant to environmental stresses.

The Value of Herd Size and Social Interaction

Goats are gregarious creatures that do best in social environments. For their behavioral and mental health, social connection is essential. Goats kept in herds are able to participate in grooming, playing, and forming social hierarchies, among other natural behaviors.

Goat welfare and behavior are also influenced by herd size. Goats can be raised in small groupings, although two or more goats is usually considered the minimum herd size. Goats that are left alone and bored can develop behavioral problems and lower productivity.

Compatibility and temperament should be taken into account while choosing herd mates in order to reduce conflict and violence within the group. Giving herd members plenty of room for exercise and enrichment activities can reduce stress and encourage wholesome social relationships.

In summary, optimal living conditions, preserving the quantity and quality of fodder, and placing a high value on socialization and herd size are all necessary for the healthy growth of goats. Goat farmers may maximize the productivity, welfare, and health of their herds by taking care of these factors, which will ultimately result in a successful and lucrative goat farming business.

CHAPTER THREE
Breeding and Reproduction

Goats' Sexual Maturity and Puberty

Goat females, referred to as does, attain sexual maturity at varying ages contingent on breed and personal growth.

While full puberty may take up to 12 months, does can reach sexual maturity as early as four months.

Goats that are male and referred to as bucks are generally ready to reproduce as early as four months of age, and they get ever more fertile until they are about four years old. In order to avoid accidental reproduction and guarantee appropriate management of breeding schedules, early separation of bucks and does is crucial.

The Ideal Age to Start a Breed

1. It is optimal for doelings, or young female goats, to breed for the first time when they are seven to ten months old.

2. Breeding at this age promotes healthier pregnancies and births by allowing for appropriate physical development.

3. Early breeding might cause difficulties and health problems for the doe and her young.

4. Pregnancy-related nutrition and care are essential for the health and welfare of the doe and her offspring.

Seasons of Breeding and Mating Systems

1. To enable spring kidding, goats are usually married once a year in the fall, which corresponds with their natural breeding season.
2. The chance of multiple births and successful pregnancies rises with fall mating.
3. Extended breeding seasons offer for more flexibility in terms of when to breed and when to kidding, which can be adjusted in response to changes in the weather and the availability of forage.
4. For highly regulated breeding regimens that guarantee consistent kidding dates for managerial objectives, synchronized breeding

may   be   utilized.
5. Breeding timing may also be influenced by considerations for ethnic or alternative markets.

Ratio of Male to Female for Appropriate Breeding

1. Depending on the breeding system being used, different male-to-female ratios are ideal.

2. One male for every 30 to 40 females is typical in traditional breeding systems where bucks are housed with does all year round.

3. A male to 20 or fewer females may be the ideal ratio for synchronized breeding, which is regulated and timed breeding.

4. By keeping the right ratio, one may minimize stress and rivalry among males and assure effective breeding.

In conclusion, successful goat farming requires an awareness of the reproductive physiology of goats, including sexual maturation, breeding age, mating systems, and male-to-female ratios. Goat farmers can optimize reproductive success and maintain the well-being and productivity of their herds by putting into practice suitable breeding procedures and management strategies.

5. Successful reproduction and general herd health are influenced by the appropriate management of breeding groups and ratios.

CHAPTER FOUR
Nutrition of Goats

Hay and Forage Are Important

Since goats are ruminants, they have a special digestive system that makes it easy for them to break down fibrous plant materials. Aside from access to grazing, goats obtain most of their nutrients from hay and fodder. These roughages give goats the vital fiber, proteins, minerals, and vitamins they need to stay healthy and productive.

Fiber: Goats require a lot of fiber for healthy digestion and rumen function, which is abundant in hay and pasture. Consuming enough fiber helps avoid digestive problems including bloating and constipation.

Protein: Specific hay and forage varieties, such legumes (like alfalfa and clover), give goats the high-quality protein they need for growth, development of their muscles, and production of milk.

Minerals and Vitamins: Hay and pasture also provide vital minerals (such calcium and phosphorus) and vitamins (like vitamin A and E) that goats need to stay healthy overall and be able to reproduce.

Appropriate Meals for Goat Farming
Supplementary feeds may be required in addition to hay and pasture to suit the unique nutritional needs of goats, particularly during times of high development, production, or gestation. Goat husbandry can profit from the following feeds:

Concentrates: Goats can get extra protein, energy, vitamins, and minerals from commercial goat diets or concentrates made especially for them. The nutritional requirements of the goats and the particular stage of production should be taken into consideration while selecting these concentrates, which can be found in pellet, grain, or mixed forms.

Grain: Goats can benefit from energy-dense feeds such as corn, oats, barley, and wheat, especially during lactation or periods of rapid growth. Nonetheless, in order to avoid weight and digestive issues, grains should be consumed in moderation.

Protein Supplements: To increase protein intake and maintain development and reproduction in situations when forage quality is low or insufficient, protein supplements such as soybean meal, cottonseed meal, or protein-rich grains (such peas and lentils) can be added to the diet.

Mineral Blocks and Supplements: Giving goats mineral blocks or loose mineral supplements made especially for them guarantees that they get enough of trace minerals like copper, zinc, and selenium, as well as important minerals like calcium, phosphorus, and magnesium.

Macronutrient Needs at Various Stages

Young goats: For healthy growth and development, young goats require a lot of nutrients. Throughout the first few weeks of life, they should be fed premium milk or milk substitute. Kids need a balanced diet high in protein and energy after weaning in order to sustain their fast growth and muscular development.

Pregnant Does: To support fetal development and get ready for nursing, pregnant women do need to consume more energy, protein, and minerals. Healthy children and adequate milk production after delivery are ensured by feeding a diet high in protein and minerals during the latter stages of pregnancy.

Lactating Does: In order to sustain milk production and the feeding of their kids, lactating does have high calorie and protein requirements. Giving them enough high-quality grass, along with grain or concentrate supplements, can help them get the extra nutrition they require during lactation.

Bucks: In order to maintain ideal physical condition and reproductive effectiveness, breeding bucks also need a balanced diet. Sufficient consumption of protein, energy, and minerals promotes sperm production and general health of the reproductive system in bucks.

It is crucial to comprehend the distinct dietary needs of goats at various phases of their lives in order to create a feeding schedule that encourages the best possible health, growth, and productivity in your goat herd. Goats are given the nourishment they require to thrive when their bodily condition is regularly monitored and their diet and feeding schedule are adjusted.

CHAPTER FIVE
Profitable Breeds of Goats for Businesses

Features of Breeds with High Meat Production

For goat husbandry to be profitable, high meat production is essential. Choosing breeds with a reputation for producing meat that is of exceptional quality will greatly improve your commercial opportunities. The following are some essential traits of breeds with high meat production:

Muscle Development: Animals from breeds with superior muscle development produce more meat overall. Seek for breeds like the Boer goat that are naturally robust and well-formed.

Quick Growth Rate: Select breeds recognized for having quick growth rates. These breeds maximize efficiency and save feeding expenditures by reaching market weight quickly.

For example, boer goats are well known for growing quickly and producing meat quickly.

Adaptability: Select breeds that do well under different conditions and with different kinds of management. Breeds that are adaptable may survive a variety of climates and grazing environments, which lowers the possibility of output losses brought on by external variables.

Disease Resistance: Take into account breeds that show resistance to typical diseases that are frequent in your area. Less veterinary care is needed for healthier animals, which lowers production costs and boosts total profitability.

Elevated Carcass Quality: Seek out breeds that yield meat with flavors, leanness, and other desirable qualities. Superior carcasses fetch higher prices in the market, which increases your profit potential.

Well-liked Breeds of Goats for Commercial Farming

Commercial farmers have come to favor a number of goat varieties because of their profitability and adaptability to different production techniques. The following are a few well-liked goat breeds for industrial farming:

Boer Goat: Native to South Africa, the Boer goat is well-known for producing superb meat. Boer goats are a popular choice for meat production enterprises globally due to their outstanding carcass outputs, high reproductive rates, and rapid growth.

Kiko Goat: Proud of their exceptional maternal instincts, resilience, and versatility, Kiko goats originated in New Zealand. Kikos are noted for producing meat effectively with little assistance from the management team and for thriving in large grazing environments.

Spanish Goat: Often referred to as brush goats or scrub goats, Spanish goats thrive in arid conditions and rough terrain. Their robust health, strong foraging skills, and high-quality meat make them perfect for low-input, sustainable agricultural systems.

Myotonic Goat (Fainting Goat): Myotonic goats have the ability to provide meat, however they are mostly kept for novelty and entertainment. When startled, these goats have a peculiar muscular rigidity that can produce tender meat with unusual flavor attributes.

Nubian Goat: Nubian goats are prized for their meat qualities in addition to their great milk production capacity. Nubians are a versatile option for dual-purpose farming operations because they yield tasty meat with good carcass conformation.

The ideal breed for your goat farming endeavor will rely on a number of variables, such as your production objectives, the state of the environment, and consumer demand. Choose the breed that best suits your goals and available resources by doing extensive study and speaking with seasoned breeders.

CHAPTER SIX
Assessing the Quality of Goats

The Outer Features of a Quality Goat

Conformation: A goat that has been bred correctly should have a balanced body structure and appropriate conformation. This involves having a balanced physique, powerful legs, and a straight back. A goat with good conformation is more capable of withstanding the hardships of farming and yielding more.

Body Type: A healthy, balanced body type is ideal for a goat; it should neither be overweight nor underweight. A healthy weight is indicated by easily felt but not bulging ribs. A healthy goat has a higher chance of thriving and yielding valuable goods like milk or meat.

Muscle: Particularly for the production of meat, muscle is a crucial component of goat quality. Strong muscles are essential for a quality goat, especially in the loin and hindquarters. Muscle enhances the overall quality of the carcass and increases meat output.

Coat and Skin: The coat should be shiny, silky, and devoid of any parasitic or disease-related symptoms. Skin that is elastic and flexible indicates adequate hydration and general health. Examining any anomalies, including sores, scabs, or hair loss, is important.

Head and Neck: The head should have alert eyes and upright ears, in proportion to the body. A robust and securely fastened neck is ideal for the human body. The shape of the goat's head and neck can reveal information about its general health and energy level.

Steer clear of common structural errors

Problems with the Legs and Feet: Deformities in these areas can significantly affect a goat's range of motion and general health. Hoof abnormalities, weak pasterns, and bent legs are common problems. Such defects may make it difficult for the goats to walk and graze effectively.

Back and Spine Alignment: Since the back supports the spine and internal organs, a robust, straight back is preferred in goats. Steer clear of goats with swayback or highly arched backs as they might cause health issues and decreased yield.

Udder Conformation: The ability of dairy goats to produce milk depends on their udder conformation. A perfect udder should have no lumps or swelling, be symmetrical, and be well-attached. Steer clear of goats with pendulous or malformed udders since they can have problems producing milk and maintaining the health of their udders.

Reproductive Health: Look for anomalies or indications of illness in the reproductive organs of breeding goats. Problems including prolapses, hernias, or undescended testicles might impact a woman's ability to conceive and reproduce.

General Health and vitality: A goat's behavior and appearance both convey its general health and vitality. Keep an eye out for symptoms of aberrant behavior, dull coat, or lethargy since these could point to underlying health issues. Select goats with an excellent appetite, digestion, and level of activity.

Through meticulous assessment of the physical characteristics and avoidance of frequent structural defects, you may choose superior goats that will better fit your farming objectives and help build a profitable goat business. Long-term productivity and profitability depend on regular health and conformation monitoring and management of goats.

CHAPTER SEVEN
The Potential Revenue from Goat Farming in Different Nations

Goat farming is a profitable enterprise that offers substantial earning possibilities in many different nations worldwide. Goat farming's profitability is influenced by a number of variables, such as consumer demand, weather patterns, agricultural practices, and governmental regulations. Here, we examine the potential revenue streams from goat husbandry in several areas:

Americas

1. Goat farming has become more popular in the US as a result of the growing demand from a variety of ethnic groups for goat milk, meat, and other products.

2. The possibility for profit differs according to the size of the business; smaller companies

concentrate on specialized markets, while larger businesses service mainstream markets.

3. The average income per goat might vary from $50 to $300 based on the market and the type of product (meat, milk, or fiber), according to USDA reports.

India

1. With a sizable market for both meat and milk, India is one of the world's top producers of goats.

2. India's income potential varies greatly depending on breed, location, and production strategy.

3. In India, commercial goat farming has the potential to yield significant profits. A single acre of land can fetch farmers anything from INR 50,000 to INR 100,000 annually.

Nigeria

1. The great demand for goat milk and meat (chevon) has led to a booming goat rearing sector in Nigeria.

2. The business is dominated by smallholder farmers, whose potential for profit varies according to breed choice, market accessibility, and management techniques.

3. Small-scale goat farmers in Nigeria can make between NGN 50,000 and NGN 200,000 year from their business; larger farms can make much more.

Australia

1. In Australia, the main goal of goat farming is to produce meat, with Boer goats being the most widely utilized breed.

2. Australia's prospective income is contingent upon various factors, including feed costs, market pricing, and export prospects.

3. In Australia, commercial goat farmers can make good money; estimates range from AUD 80 to AUD 150 for each goat sold for meat.

Kenya

1. The rising demand for goat milk and meat in urban areas has led to a growth in the goat farming sector in Kenya.

2. Kenyan income potential varies according on breed, market accessibility, and productivity levels.

3. According to certain sources, small-scale goat farmers in Kenya can make a substantial annual revenue per acre of land, ranging from KES 30,000 to KES 100,000.

Brazil

1. Brazil is a leading producer of goat meat, emphasizing large-scale grazing operations in areas that are semi-arid.

2. Brazil's potential income is influenced by a number of variables, including market pricing, productivity, and herd size.

3. According to some estimates, commercial goat farmers in Brazil can make between BRL 50 and BRL 150 for each goat sold for meat.

All things considered, goat farming presents a positive financial outlook for producers, both commercial and small-scale, throughout a number of nations. To achieve maximum profitability, goat farming involves effective planning, management, and market access.

CHAPTER EIGHT
The benefits and drawbacks of the goat farming industry

Benefits

High Meat and Milk Production: Raising goats has a number of benefits for producing a lot of meat and milk. Goats are prolific breeders and can produce large amounts of high-quality meat and milk if given the right care. Mutton, also referred to as chevon, is a tasty, lean meat that is highly sought after around the world. Goat milk is also very nutrient-dense, quickly absorbed, and safe for people who are lactose intolerant.

Adaptability to Different habitats: Goats' exceptional ness to adjust to a wide range of habitats and climatic conditions is one of their most notable traits. In hostile environments, such as desert and semi-arid zones, where other cattle might find it difficult to survive, they can flourish. Goat farming may be done in a variety of geographical settings because to its versatility, which gives farmers freedom in selecting their agricultural grounds.

Profitability Potential: There is a lot of room for profit in goat farming because of a number of things. First off, compared to larger livestock species like cattle, goats require a very little initial investment. They also attain market weight more quickly and have a shorter gestation time, which facilitates faster turnover and possible money creation. Furthermore, there are potential for farmers to profit from market demand as goat meat and milk demand is growing both locally and globally.

Drawbacks

Health Issues and Disease Risks: The incidence of health issues and disease risks is one of the main worries in goat husbandry. Goats' productivity and profitability can be greatly impacted by a number of viral and parasitic disorders, including pneumonia, foot rot, and internal parasites. Goat farming might incur higher operating costs and labor due to the need for strict hygienic practices, vaccination schedules, and routine veterinarian care in order to manage and avoid infections.

Infrastructure and Initial Investment Requirements: Starting a goat farming enterprise involves a sizable upfront investment in resources and infrastructure. This include finding appropriate land for habitation or grazing, building shelters, setting up fences, and buying supplies and equipment. Furthermore, higher genetic quality breeding stock may come with higher prices. Enough facilities and infrastructure are necessary to guarantee the health and productivity of the goats, but they might be too expensive for aspiring farmers, especially those with little money.

Market Volatility and Competition: These factors might have an impact on sustainability and profitability in the goat farming sector. The financial sustainability of goat farming enterprises can be impacted by shifts in customer tastes and fluctuations in the market prices for goat milk and meat. In addition, pricing and market dynamics may be impacted by imports of goat products and competition from other livestock producers. To counteract the consequences of competition and market instability, farmers must diversify their product offerings, adapt to market trends, and look into niche markets.

In conclusion, goat farming has many benefits, including high yields of meat and milk, flexibility, and potential for financial gain, but it also has drawbacks, including health problems, the need for a large initial investment, and unpredictability in the market. Careful planning, proactive management, and innovative approaches are necessary for goat farming enterprises to succeed in this fast-paced industry and overcome obstacles.

CHAPTER NINE
Elements Influencing the Production of Goats

Grazing Management and Stocking Rate

Definition: The quantity of animals grazing on a specific plot of land for a predetermined amount of time is known as the stocking rate. It is essential to figuring out how productive and long-lasting goat farming enterprises are.

Relevance:

a. Overgrazing, soil erosion, and the depletion of pasture supplies are all consequences of overstocking that can lower productivity.

b. Understocking could lead to underuse of the forage that is available, missing out on chances to maximize output.

Grazing Management Techniques

Rotational grazing: To allow fodder recovery, pasture is divided into smaller paddocks and goats are rotated between them.

Rest Periods: To sustain long-term productivity, pastures should be given time to recuperate and rest following grazing.

Supplemental Feeding: Giving animals extra feed when there isn't much grass available or when pasture conditions aren't ideal.

Plant   Species   Composition's   Effect
Diversity of Forage

a. Goats can choose from a variety of plant species that have differing nutritional profiles, digestibility, and palatability.

b. A balanced diet can be produced by a variety of forage species, which lowers the chance of nutrient deficiencies and improves production and health.

Adjustment to Regional Circumstances

a. A variety of forage species, including as grasses, legumes, and browse plants, are suitable for goat health.

b. The choice of forage plants that are compatible with the soil, climate, and management techniques of the area can maximize resilience and productivity in goat farming   systems.

Quantity and Quality of Forage

a. Goat development, reproduction, and milk production are impacted by the nutritional content and availability of fodder, which is determined by the species mix of plants.

b. The productivity and composition of pasture vegetation can be impacted by management techniques such weed control, irrigation, and fertilization.

Animal Genotype and Production Stage

Variations in Genetics

a. The characteristics of goat breeds differ in terms of growth rate, milk yield, resistance to disease, and environmental suitability.

b. Performance and profitability can be maximized by choosing breeds or genetic lines that are compatible with particular production objectives and environmental circumstances.

Production Stage

a. Goats require different nutrition during different physiological stages, such as growth, lactation, pregnancy, and maintenance.

b. To maximize productivity and health, management techniques should be adjusted to the changing dietary requirements of goats during their lifecycle.

Reproduction and Health

a. In goats, genetic variables affect reproductive success as well as vulnerability to illnesses and parasites.

b. Genetic qualities linked to overall productivity, reproductive efficiency, and disease resistance should be given priority in breeding selection and management procedures.

In summary, optimizing productivity, sustainability, and profitability in goat farming operations requires careful management of the stocking rate, grazing techniques, forage species composition, and genetic selection. Goat farmers may maximize resource use, improve animal health and wellbeing, and meet production targets by being aware of and taking action against these concerns.

CHAPTER TEN
Goat milk-Benefits and Drawbacks

Advantages of Drinking Goat Milk

Since ancient times, people have valued goat milk for its many health advantages. The following are a few benefits of eating goat milk:

Digestibility: Because goat milk contains smaller fat globules and a different protein composition than cow's milk, it is easier to digest. For those who have dairy sensitivity or lactose intolerance, this makes it a good substitute.

Nutritional Value: Vitamins A and B, calcium, phosphorus, and potassium are among the many important components found in goat milk. It is a nutrient-dense option for general health and wellbeing because it also has higher concentrations of several minerals than cow's milk.

Lower Allergenicity: Because goat milk has a different protein composition than cow's milk, some persons who are allergic to cow's milk may be able to take it better. For those who are susceptible, goat milk's reduced alpha-S1-casein protein content may mean less allergic reactions.

Digestive Health: Goat milk's oligosaccharides may encourage the growth of advantageous gut bacteria, promoting the health of the digestive system and enhancing nutrient absorption.

Boosts Immunity: The antibacterial qualities of bioactive substances like lactoferrin and lysozyme found in goat milk may aid in fortifying the immune system.

Skin Health: Goat milk is a great element in skincare products because of its high fat content, which hydrates the skin. Additionally, it might lessen the symptoms of skin disorders including psoriasis and eczema.

Possible Allergic Reactions and Side Effects
Although goat milk has many health advantages, there are risks associated with it for some

people, especially those who have certain allergies or sensitivities. Among the possible adverse effects and things to think about are:

Allergy Reactions: Goat milk is typically thought to be less allergenic than cow's milk, although allergy sufferers may still have allergic reactions. Casein and whey allergies in goat milk can result in symptoms ranging from moderate gastrointestinal distress to life-threatening anaphylaxis.

Cross-Reactivity: Individuals with cow's milk protein allergies may also have cross-reactive allergy reactions if they are sensitive to comparable proteins in goat milk.

Environmental Factors: If environmental pollutants like pesticides, antibiotics, or poisons are not effectively handled, they can build up in goat milk and have a negative impact on consumers' health.

Nutrient Variability: A goat's diet, breed, and lactation stage can all affect the amount of nutrients it contains. This fluctuation could have an impact on the consistency of the nutritional advantages of drinking goat milk.

Digestive Problems: Although many people find that goat milk is simpler to digest than cow's milk, some people may still have digestive problems, such as gas, bloating, or diarrhea, particularly if they have underlying gastrointestinal diseases.

Interactions between Medication and Goat Milk: Some drugs may have interactions with components in goat milk that impair their effectiveness or absorption. People who take medication should definitely speak with healthcare providers before adding goat milk to their diet.

In conclusion, even though goat milk has many health advantages and is a good substitute for people who are allergic to dairy, it's important to think about possible allergies and side effects. To find out if goat milk is a good choice for someone with particular dietary needs or restrictions, speak with a healthcare professional or allergist.

CHAPTER ELEVEN
Handling Techniques for Juvenile Goats

Age of Mother-Child Separation

It's critical for children's growth and general wellbeing to separate from their mothers at the appropriate time. Although goats in the wild naturally wean themselves by the time they are 3 to 4 months old, in domestic settings it is typically advised to hold off on separating young goats from their mothers until they are at least 2 to 3 months old. This gives the children enough time to learn how to socialize, eat solid food, and become independent of their moms. If you separate them too soon, it can create stress and nutritional inadequacies; if you separate them too late, it can interfere with the dam's reproductive cycle and result in dependency problems for the children.

Giving Baby Goats Solid Food

A crucial step in preparing newborn goats, sometimes called kids, for independent feeding is to introduce solid food to them. For the first few weeks of life, babies still mostly depend on milk to meet their nutritional needs, but as their digestive systems improve and their energy needs rise, solid food is progressively introduced. When they are 2 to 3 weeks old, start by giving them modest amounts of premium hay, fresh grass, or special baby pellets. Make sure there are no pollutants and that the food is easily digested. When they get older, gradually wean them off of milk while increasing the amount of solid food they eat.

Goats' Reproductive Age and Pregnancy

Maintaining an effective breeding program and optimizing output in goats requires an understanding of reproductive age and pregnancy management. The normal age at which female goats, or does, attain reproductive maturity is 7 to 10 months, though this can vary according on breed and individual development.

To guarantee a good pregnancy and healthy offspring, it is generally advised to wait until they are at least 60% of their adult weight before breeding. Male goats known as "bucks" are fertile as early as four months of age, but in order to guarantee superior sperm quality and reproductive efficiency, they are typically not utilized for breeding until they are older.

Goats have a 150-day gestation period, or roughly five months, after they are bred. Pregnancy requires careful attention to ensure the health and welfare of the doe and her progeny, including proper diet, veterinary treatment, and monitoring. In addition to proper postnatal care, a clean and comfortable kidding area lowers the risk of problems and increases the likelihood that newborns will be successfully raised.

Goat farmers can ensure the well-being, growth, and productivity of their herd while promoting sustainable and profitable farming practices by putting into practice appropriate management practices for young goats. These practices include timely separation from mothers, gradual introduction of solid food, and careful management of breeding and pregnancy

CHAPTER TWELVE
Final Thoughts

An overview of the main ideas

This extensive manual on goat farming success has emphasized a number of important points:

1. Raising and rearing domestic goats is known as "goat farming," and it may be a fulfilling aspect of animal husbandry.

2. Goats require proper management and care to be healthy, grow, and produce.

3. The keys to success include choosing well-bred goats, giving them enough food, and keeping their living conditions ideal.

4. Using efficient breeding techniques and comprehending the reproductive cycle are important components of sustainable goat husbandry.

5. One way to optimize income potential is to choose profitable goat breeds that align with particular business objectives.

6. Being aware of the benefits and drawbacks of goat farming aids in decision-making.

7. A number of factors, including genetics, diet, and stocking rate, have a big influence on goat output and need to be carefully controlled.

8. Goat milk has nutritional advantages, but its possible negative effects should be taken into account before consuming.

9. Young goats require proper management techniques to ensure their proper development, including when to separate from their mothers and introduce solid food.

The Value of Care and Management in Goat Farming

Goat husbandry requires careful management and attention for a number of reasons.

Health and Welfare: Providing goats with proper food, medical attention, and housing conditions enhances their overall health and well-being and lowers their susceptibility to illnesses.

Productivity and profitability: Fertile goats that are well-managed produce larger amounts of meat, milk, or fiber, which increases farmers' profits.

Genetics and Breeding: Quality genetics and strategic breeding techniques can enhance herd performance and provide progeny with desired features, so bolstering the farm's sustainability and prosperity.

Sustainable Farming Practices: By maximizing resource use, reducing environmental impact, and building resilience to obstacles like climate unpredictability, effective management practices support sustainable farming.

Reputation and Branding: Goat farms with a strong emphasis on good management and care of their animals tend to gain a favorable reputation with customers, which increases the marketability and brand value of their products.

In summary, careful management and care techniques that put the health, welfare, and productivity of the animals first are essential to the success of goat farming. Goat producers may contribute to the expanding goat farming business and achieve long-term success by putting good management methods into place and continually improving processes.

www.ingramcontent.com/pod-product-compliance
Lightning Source LLC
Chambersburg PA
CBHW072257310526
45795CB00012B/1705